Finding the Answers: Discovering World War II Service Online

Written by Jennifer Holik

WWII Research and Writing Center Publishing

Editors: Sarah Ferguson Potter and Johan van Waart
Cover Designer: Jennifer Holik

Holik, Jennifer, 1973 –
    Finding the Answers: Discovering World War II Research
Online / Jennifer Holik. Includes bibliographical references and indexes.

ISBN: 978-1-938226-48-9
ISBN: 1-938226-48-8

Also By Jennifer Holik

Finding the Answers: Researching World War II Army Service Part 1

Finding the Answers: Researching World War II Army Service Part 2

Finding the Answers: Researching Women in World War II

Finding the Answers: World War II Travel in Europe

Finding the Answers in the Individual Deceased Personnel File

Finding the Answers: Discovering World War II Service Online

Faces of War Researching Your Adopted Soldier

Stories from the World War II Battlefield
World War II Writing Prompts

Stories from the World War II Battlefield Volume 3
Writing the Stories of War

Stories from the World War II Battlefield Volume 2
Navigating the Service Records for the Navy, Coast Guard, Marine
Corps, and Merchant Marines

Stories from the World War II Battlefield Volume 1
Reconstructing Army, Air Corps, and National Guard Service

Stories from the Battlefield:
A Beginning Guide to World War II Research

The Tiger's Widow

Stories of the Lost

Engaging the Next Generation:

A Guide for Genealogy Societies and Libraries

Branching Out: Genealogy for Adults

Branching Out: Genealogy for High School Students

Branching Out: Genealogy for 4th-8th Grades Students

Branching Out: Genealogy for 1st-3rd Grade Students

To Soar with the Tigers

Dedication

For all those preserving the stories of our WWII service members.

Acknowledgements

I could not have written this book without the support of my husband Johan and my three amazing boys, Andrew, Luke, and Tyler. Thank you for always believing in me and listening to me talk through research problems. I love you all.

Virginia Davis for her constant love, encouragement, and support as I wrote all my World War II books.

Over the last five years, many amazing and knowledgeable researchers have contributed to the knowledge I have today. They include, Thulai van Maanen, Norm Richards, Jonathan Webb Deiss, Mike Constandy, Sarah Ferguson Potter, Mary Hoyer, Connie Yen, Lisa Alzo, Eric Bijtelaar, Doug Mitchell, Tom Scholtes, Vincent Orrière, Ryan Kelly, Mikel Shilling, Herman Wolters, and Sebastiaan Vonk.

A big thank you to all my clients. Through every project I expanded my knowledge which allows me to share more with others. Thank you to everyone who attended my military lectures and told me their stories or asked questions. I cannot forget the individuals who emailed me or chatted with me on social media about their soldier's story. I learned something from each of those encounters which helped create my programs and books.

Table of Contents

Introduction

Congratulations with starting your World War II research online!

Each day new materials are digitized and placed online for every imaginable subject. Due to more materials for World War II research becoming available online through subscription and free sites, many people assume everything should be available online. And that they should be available now.

This is not realistic though for many reasons. Primarily due to the cost of digitizing materials and storage of those materials. Additionally, privacy and classification are other reasons many of the records we need for military research will not be available online anytime soon.

One example of a record you will almost never find online is the Official Military Service File (OMPF), also known as the service file. The only copy of an OMPF is held at the National Personnel Records Center (NPRC) in St. Louis, Missouri.

There are also records which have not been declassified by the U.S. government. Much of what researchers require for World War II research has become available. However, if you move through Korean War and Vietnam War or into more current wars, those records are not as readily available. Finally, as you work through the material provided in this book, please keep in mind the same search tips apply to other wars you may be researching.

This guide is meant to be a starting point for World War II research, not an exhaustive examination of all the military branches and records available.

For an in-depth look at research, visit http://wwiieducation.com for educational articles, live programs, links to online courses, and the in-depth WWII research guides which include,

Stories from the World War II Battlefield

    Volume 1: Reconstructing Army, Air Corps, and National Guard

Service.

Volume 2: Navigating Service Records for the Navy, Coast Guard, Marine Corps, and Merchant Marines.

Volume 3: Writing the Stories of War.

Researching your service member's history can be complex. The World War II Research and Writing Center provides research experts to tackle your most challenging research questions. Contact us at info@wwiirwc.com for project details and costs. We have researchers on-site at the National Archives facilities to obtain materials within a few weeks, and a network of researchers and tour guides around the globe.

Why should you research online?

Why we should research World War II service online may require no explanation, because to most of us, it seems the most logical starting point. However, there are several points to consider that may influence where and how you search for information online.

You will gain an education.

Many websites, blogs, and expert forums, provide information on World War II research, discuss the process, so called brick walls, and tell the stories of service members. There is not one site or book that can provide all the answers. It is important to explore as many sources as possible.

World War II Associations have digitized records.

Many World War II units formed reunion or association groups after the war to stay connected. Over time, materials from the members were donated to the organization to establish an archive. When records were declassified years later, many organization historians began obtaining photocopies of unit records.

As technology advanced, unit historians digitized some materials for their unit and placed them on their organization websites. It is rare to find all materials for a unit available online. The 7th Armored Division and a few Bomb Groups seem to have come close. You will not however, find the OMPFs online, even within all their digitized materials. Much of this is due to past privacy and access restrictions and the 1973 Fire at the National Personnel Records Center (NPRC) in St. Louis, Missouri.

Not every unit has had the funds, technology know-how, or people, to digitize materials. The 104th Infantry Division (Timberwolves), as an example, is one association that does not have materials available on a website. However, if you contact their historian, or ask questions in their Facebook group, you will often find someone who has the records you seek.

Finally, there are some units like the 90th Infantry Division Association, which have some unit records digitized online, but their company level records are only available in paper format through their historian.

Regardless of which situation you find yourself as you research, always ask the unit association staff questions. You never know what they have available that would benefit your research.

You will often add facts to your timeline.

Many clues to military service can be found online and should be added to your timeline of service. A timeline of service is a listing by date of all important facts and events about a service member's history. To learn more about this, see the book, *Finding the Answers: Starting World War II Research,* available on the WWII Education (http://wwiieducation.com) website.

I encourage researchers to keep a list of the websites they visit and on which they locate information. Documenting this in your timeline of service allows you to refer to the website in the future and provides a source should you decide to publish your service member's story later.

Books have been digitized.

Many of the unit histories written by the units, using official military records, after World War II ended, have been digitized and made available on sites like GoogleBooks, Fold3.com, and Internet Archive. Sites like these are one of the first places you should look for information.

Connect with other researchers and veterans.

Collaborating. Networking. Sharing knowledge: these are valuable assets for expanding military knowledge.

In 2010, when I began my investigation into the research methods for all military branches in World War II, I made connections through Facebook with researchers in Europe and the Pacific. It wasn't long before I got to know people researching specific units, battles, locations, or focusing on those who never returned from war.

Connecting with these researchers enabled us to share our knowledge, grow our research archives, and network with other people looking for information. In addition, these are the experts I call upon when I travel in Europe or have a client who wishes to walk in their father's footsteps. The researchers in my network have boots on the ground knowledge of the battlefields, local museums, and points of interest, that you will usually not learn about on a big tour.

Finally, when I have a question I cannot seem to answer, I can call upon any of these people for help. If they do not have an answer, chances are great that they know someone who does.

Connect with family.

There are two family connections available to most researchers. First, we have our biological families, for which many online family trees exist. Be aware that many have inaccurate information copied from other people's trees. Check your facts before you accept what you discover as truth.

Second, there are many people in Europe who have adopted the grave of a World War I or World War II service member in the American Battle Monuments Commission (ABMC) cemeteries. Many Europeans see their service member as a member of their biological family. They honor the memory of those buried overseas. These grave adopters are always trying to connect with the biological family of the service member, whose grave they adopted. Many grave adopters have information about the service member that is unknown to biological families. When you can connect with these researchers, I encourage you to do so.

Online Research Techniques

Prior to starting online research, I encourage you to write down what you know about your service member and create a timeline of service. Include a source for each fact you write down so you know where to look when you find conflicting information. Errors can occur through our transcription of information from one source to our notes. They may occur within the military records or information created for websites. Documenting where you found the information for each fact will save you time down the road. Check and double check your facts.

Document the websites you visit.

Document the websites you visit and keep track of what information you located there. Websites change content or disappear over time. Knowing what you discovered when, and where, will help keep your research organized. To help you with this task, you can use the Online Research Summery sheet at the end of this chapter.

Check the date of information.

It's important to be aware of the "update date" on a website where you find information on how to research. Some of what circulates on the internet about records access for World War II records is inaccurate. Many website administrators are either unaware and do not update their sites or once the site is up, nothing is ever updated. For instance, any site that tells you to write to College Park, MD for an Individual Deceased Personnel File (IDPF) is out of date. The same applies to any site that tells you only next-of-kin can access World War II OMPFs. Records access for these records has changed in the last few years.

Use indexes and understand the limitations.

Indexes for online records will only get you so far. Be aware that indexing errors occur when records are difficult to read, often due to age, wear and tear, fire, or water damage. In addition, when records are handwritten, the way one person deciphers the text may not be the same way you would. Consult other sources when possible if you believe the information you seek should be in the index, and is not.

Indexing may not capture every record. The USMC Muster Rolls which are digitized on Ancestry.com are a good example. While the Muster Rolls for every month exist on Ancestry, the index will only pull up results for the months January, April, July, and October. I am not sure why the indexing was only done quarterly, but this is all that is available when you search the index for a Marine.

How do you locate information if it exists but is not indexed? As an example, if you search Ancestry for Marine William F. Cowart, he appears on the USMC Muster Rolls in January, April, and October 1943. However, he does not appear in the index in November, the month he was killed on Tarawa.

At the top of each page on Ancestry, I pulled up for Cowart, there is a dropdown listing the Roll number on which that page exists. Knowing his unit, I was able to move forward through the roll numbers to locate the November 1943 Muster Rolls. Conducting a page by page search, I was able to locate his Killed In Action (KIA) entry for November 1943.

The USMC Muster Rolls may not be the only digitized record set that was indexed this way. Be aware that most rosters were compiled at least on a monthly basis. What you see after searching the index, may not be all that is available.

Search the collaterals.

Using online resources, you can search for the collaterals. Collaterals are those individuals who served with your service member. Locating pieces of their stories may help you tell the story of your service member.

It may be unreasonable to search for every infantryman who served with your Army soldier, so instead, look for the men who ran the company. If your soldier was part of a bomb crew, the list of those he served with will be much shorter.

For example, several years ago I researched the service for 2nd Lt. Fred A. Davis, who was the co-pilot on a bomber that went down 2 November 1943 in Austria. Searching online yielded little for Fred.

Searching for the pilot of the bomber however, yielded a pot of gold I could not have imagined!

I located a website dedicated to the mission on which Fred's plane was shot down. The website author had scoured American, Austrian, and German archives and written an 800+ page book about the mission and all the crews. The book even had photographs. The author never published the book, he said due to not having all the rights to the photos. He did share it with me on the condition I didn't share it with others. His work was so well documented, it would be relatively easy to retrace his steps to verify any fact he listed. His book also gave me a photo of Fred in his bomb crew.

You never know what you will find by searching the collaterals!

Locating Information Using a Search Engine

To locate information on a specific individual or unit, I have listed many techniques. Try each one using your favorite search engine. Each will return slightly different results based on how that engine categorizes websites. We will use Pvt. James Privoznik as an example for our search.

Known Details:

U.S. Army

Pvt. James Privoznik, 36640529 from Illinois

Final unit: "F" Company, 358th Infantry Regiment, 90th Infantry Division

KIA 11 January 1945

- Use quotes around exact phrases you wish to search. "James Privoznik" is an example.

- Search by name and serial/service number. Try James Privoznik 36640529, or just the number. Usually the results will come from the NARA Enlistment Database, if the soldier enlisted in the Army. Sometimes it will come from articles or blog posts, unless records have been indexed.

• Use specific and unique terms. Try 90th Infantry Division, 358th Infantry WWII, 90th Division WWII, or any combination.

• Capitalization usually doesn't matter when searching.

• Try the wildcard using the * symbol. WWII* or Privoznik*

• Change your search preferences to search a specific date range of items posted online. For example, maybe you are looking for a person and only articles posted this year or a specific date range of 2012-2014.

• Try using advanced searches to add more criteria to the search. ProQuest newspaper search is one engine that uses this function.

• Search for the name of a group using different spellings. For example, the 100th Bomb Group was called the Bloody Hundredth or Bloody 100th. Searching all three options may provide different search results.

• For Army Air Forces, try searching for the name or number of the plane flown, the name of a pilot, or names of bomb crew members.

• Search for names of bridges taken, battles fought, cities bombed, specific Hills (and their numbers).

• Try a specific group and the name of a military report you wish to locate. For example, 100th Bomb Group Mission Report, 90th Division After Action Report, or 327 Engineer Morning Reports.

• Creatively search the results that appear in a search. Digitized materials are prepared by Optical Character Recognition (OCR.) OCR picks up approximately 80% of the words in a scanned document, which leaves a lot of room for researchers to miss key records. This happens more in military documents which are blurry, damaged from weather, fire, water, or other reasons, or were not in great condition when they were originally scanned. Not all military records, regardless of archive from which they come, look like they were just printed off a laser printer.

Most Commonly Asked Question

A question I am often asked by researchers is, 'How do I find the unit my service member belonged to if I don't have his papers?' Honestly, this is a hit or miss search. If documents with the soldier's name have been digitized or written about in an article or blog post, you may locate them through a search.

Online Research Summary

Date	Name of website with URL	Notes

A Sample of Websites with World War II Record Collections or Information

Every day new websites are added with World War II information. It is impossible to list every website I have ever used and found helpful. I can touch on a few of the larger websites which contain digitized records. Websites continually update, so it is a good idea to visit often to see what's new. Some offer newsletter subscriptions which notify subscribers when new information has been added.

Often we start our searches surname or a combination of surname plus given name. Depending on the search criteria available, adding date of birth, death, or date of service may narrow down search results.

The links to the sites listed in this book can also be found in the Research section of the World War II Research and Writing Center's website. http://wwiirwc.com.

General Resources
American Battle Monuments Commission (ABMC)
http://abmc.gov

This is a free website created by the federal government. It contains a database listing all service members buried in the World War I and World War II American Battle Monument Commission (ABMC) American Military Cemeteries overseas, or listed on the Tablets of the Missing. The site also includes histories of the cemeteries and educational resources.

Search Tips: From the main page, you can search by surname. Choosing the link that says Search ABMC Burials, allows a more detailed search in which you can include surname, filter them by war, cemetery (if known), or branch of the service, in addition to other options.

Ancestry.com
http://ancestry.com

Ancestry.com is a subscription based website to help researchers create online family trees and locate genealogical information. The site contains many military databases, some of which can be found free on other websites. Many libraries offer cardholders free access to Ancestry.com for use onsite.

Search Tips: Begin your search with the soldier's name and date of birth. Explore the military resources which appear to see if there is a match. Explore the online family trees to see if your soldier has been entered and if there is additional information you do not have. Contact the tree owner if you believe there is a match within a private tree. You can also try searching the card catalog for specific military databases and searching within those.

Digital Public Library of America
http://dp.la

The Digital Public Library of America brings together media resources from libraries across the country in one place.

Search Tips: Search by general keywords like "World War II," or be more specific with "82nd Airborne" and see what results appear. The results screen allows you to filter the results by media type, along the left side of the screen. Books may take you to the Hathi Trust website, where you can view the books, but not download unless you have a partner login. Even if you cannot download the book, if you have enough interest in it, obtain it through interlibrary loan, or see if Internet Archive has a copy.

FamilySearch
http://familysearch.org

FamilySearch is a free website run by the Mormon Church. You can create free family trees, search for records and read digitized books. A microfilm service used to be available to local Family History

Libraries. This service has been discontinued as many of those records are now available on their website. In cases where access is restricted, some records can be viewed only at a Family History Library. The main record source available on FamilySearch is the World War II Draft Registration Cards, but there are a few state-level record groups.

Search Tips: I always find it easier to search by area in the world and within that area, a specific record set. This focuses my search and does not bring up every possible record within FamilySearch's enormous database of records. How do you search like this?

From the main page click the SEARCH icon. When the map appears, select your location by continent. Then choose the state you wish to view, and the record set within the state. Not all FamilySearch images have been indexed. There is a good chance you will need to sort through many images of records to locate an index, which may exist, within a record set.

FamilySearch Digitized Books
http://bit.ly/1xGAnrl

FamilySearch has not only digitized genealogical records from all over the world, but also has access to more than 150,000 books through its own library and partner libraries.

Search Tips: Use the Search function to locate books with the subject World War II. The book results can be sorted by material type, collection, language, and author/creator. The search results also provide additional subject related to World War II you can view along the left sidebar.

Fields of Honor Database
http://fieldsofhonor-database.com/

The Fields of Honor Database is a compilation of all the names of service members buried and listed on the Walls of the Missing at several cemeteries in Belgium, France, Luxembourg, and the Netherlands. The database contains information on service and when available, a photograph of the service members. Sources for the information provided are cited at the bottom of the entry. Typically these are online sources only, though the Fields of Honor volunteers do have military records in their archives. The Fields of Honor volunteers respond to questions and accept photographs and information on soldiers in the database.

Search Tips: Start by searching for the soldier by name. The database can be a little finicky so be sure to capitalize the surname and name. Rosenkrantz, David. You can also select a specific cemetery and go to the letter of the surname to view all those in the database.

FindAGrave
http://findagrave.com

FindAGrave is a free, volunteer run site containing grave listings, often with photos. One issue with this site is if there are multiple entries for the same individual at the same cemetery, they may be linked as siblings rather than stand-alone.

Search Tips: Start by searching the 132+ million grave records with the link on the right side of the page. This link brings up a page with many search options from name, dates of birth and death, grave location, and filters. Please remember the ABMC created a memorial for each individual buried at their cemeteries or listed on the Tablets of the Missing. Others may have also created memorials for these same people. Use caution when gathering this information and pay attention to IF or WHERE the individual is actually buried. It could be they are still Missing In Action or just have a Memorial Stone.

Fold3.com
http://fold3.com

Fold3.com is a paid subscription site owned by Ancestry.com, offering primarily military unit records, newspapers, and some genealogical records. In some cases you will discover company or squadron level records, usually if a reunion association donated the materials for the Fold3.com archive. Like the Ancestry.com subscription, many libraries offer card holders free access to this database on-site at the library.

The databases can be a little difficult to navigate if you only search by name. Usually the information available to tell your service member's story will not be discovered searching with name only.

Search Tips: The records on Fold3.com are usually unit records which may more often list officer's names rather than enlisted men. The unit records provide more of a historical context examination of the war. Reports were created for specific missions, battles or campaigns, analysis of action that took place, statistics for a unit, and other higher level details. While your service member may not be listed specifically in these reports, reading them explains his role and his unit's role in the entire war effort. This historical context is valuable for his overall story.

Start by searching for the soldier's name in the WWII Record collections from the main search screen. For an example, let's look for records for Staff Sergeant David Rosenkrantz of I Company, 504th PIR, 82nd Airborne. PIR stands for Parachute Infantry Regiment, a phrase which may or may not be spelled out in records. Searching for David's name does not bring up many results, although we see many of the same records as can be found on Ancestry.com.

Since we know David was in the 504th PIR, we can go directly to specific record sets to search. Go into Modify Search and remove David's name, but enter 504th into the Keyword box. This is where searching takes on a whole new dimension. You have to play with 504th versus 504 and look for records that show 504th PIR or 82nd Airborne. When we search for 504th, we are hoping to stumble upon an entry about David, but also looking for historical context and battle information.

Two specific record sets to search, in which you can usually find a lot of information, are WWII European Theater Army Records and WWII Foreign Military Studies, 1945-1954. Searching within these sets for units or battles often brings great results.

The key to locating information in Fold3.com is to be patient and try every possibility from keywords, names, and searching within record sets. Finally, do not limit the end date in the data range. Many analysis reports were written in 1946 and later which provide analysis of battles and units.

Google Books
http://books.google.com/

Google Books is a collection of digitized public domain books and book previews. It is a great resource for out of print books or discovering a source to research.

Search Tips: Search by book title or author or general keywords to see what books are available. If a book you suspect should be in the public domain appears on the list, but not in full text availability, see if it exists on Internet Archive.

Historical Newspaper Archives
There are several historical newspaper archives available in the U.S. I will not attempt to list them all in this chapter, but will provide a few sites where you can begin a search. Also check ProQuest, Ancestry.com, and Fold3.com, which are discussed in this chapter.

Chronicling America
http://chroniclingamerica.loc.gov/

The Library of Congress newspaper archives.

Genealogy Bank
http://www.genealogybank.com/

A subscription site with more than 7,000 newspapers.

Newspaper Archive
http://newspaperarchive.com/

A subscription site with more than 400 years of newspapers for all 50 U.S. states and 22 other countries.

Search Tip: Search for newspapers within the locale your soldier lived or where he trained. You may find free online newspapers or libraries which offer free look-ups. Check local archives in Europe for newspapers which may provide historical context and photographs for the unit in which your soldier served.

Internet Archive
https://archive.org/

Internet Archive is a digital library which houses public domain books, manuscripts, music, videos, and photographs.

Search Tips: Search Internet Archive for media specific items. For example, start by choosing books and search within that only for 82nd Airborne. What books do you discover? Now try the same search in all media. What appears?

Library of Congress
http://loc.gov

The Library of Congress is a free resource which covers many topics of interest to the people of the United States. Veterans Oral Histories can be searched here, and you can learn how to submit your own. There are resources for newspapers, maps, and other digitized materials related to World War II.

Search Tips: The Library of Congress (LOC) website has an easy to use search function which allows you to conduct a general search or a specific media search. On the results page, along the left side of the screen, are several categories by which you can filter the results.

National Archives
http://archives.gov

The National Archives is a repository for federal records. Online databases include:
• Electronic Army Serial Number Merged File, ca. 1938 - 1946 (Enlistment Records)
• Electronic Army Serial Number Merged File, ca. 1938 - 1946 (Reserve Corps Records)
• World War II Prisoners of War Data File, 12/7/1941 - 11/19/1946
• World War II Prisoners of the Japanese File, 2007 Update, ca. 1941 - ca. 1945
• Japanese-American Internee Data File, 1942 - 1946
• Naval Group China Muster Roll and Report of Change Punch Cards, 1942 – 1945

Search Tips: Search the NARA databases by surname, given name. If you have a common surname and the search form allows for an advanced search, try different variables in the search. Most of the database you find on NARA's website are also on Ancestry.com and Fold3.com.

Pritzker Military Museum and Library
http://pritzkermilitary.org

The Pritzker Military Museum and Library is a non-partisan research library focusing primarily on the stories of the Citizen Soldier through all wars. The library houses books on many military topics including conflicts, histories, equipment, and individual soldier stories. Books are available in English and several other languages, which allow researchers to learn different views of a war or the outcome.

Search Tips: The library website is easy to navigate. When searching for books, which you can request through interlibrary loan, search by author, keyword, or title. Other media options are available on the results screen along the right side of the screen.

ProQuest
http://proquest.com/

ProQuest is a subscription based service offered through most libraries, and is usually accessible from home using your library card. Newspapers and databases vary based on location and what databases the library subscribes to each year. For example, in Chicago libraries, most ProQuest services include the Historical Chicago Tribune, Historical New York Times, and a variety of database.

Search Tips: I use ProQuest a lot for newspaper research. It is, again, a resource which has OCR and only 80% of the text appears in search results. There are times when you need to do a page by page or day by day search for specific articles if you cannot find them through search.

For example, during World War II, the newspaper printed articles, often titled, "Army Dead," on average, three times a week. Those articles contained the names of soldiers Missing, Wounded, Prisoner, or Killed In Action with the name of their next-of-kin and home address. These lists were printed 30-90 days, usually, after the War Department notified the family of the loss or change in status. I have had to use the search daily approach to find the names of cousins who were labeled KIA during the war. It requires a bit of patience.

White Pages – Online Phone Directories
http://www.whitepages.com/

Several online phone directories exist, and White Pages is only one option. These directories can be used to locate next-of-kin if you have a location nailed down where they possibly live.

Search Tip: Google "White Pages" or "Phone Directory" for a general search, or add the locale to the search to narrow down your options.

Wikipedia
http://wikipedia.org

Wikipedia is an online encyclopedia of information created by various contributors. It almost always appears on the first page of search engine results. While Wikipedia can provide some background information, and may provide a source or two for you to investigate, it should be used with caution.

Example: 35th Infantry Division (United States)
https://en.wikipedia.org/wiki/35th_Infantry_Division_(United_States)

This article contains a lot of useful information to be used as background information. Much of it is sourced if you look at the bottom of the page. This allows you to locate and review the sources used. The article could benefit from improvements, which would fill out all the units attached or which belonged to the 35th so you could learn more about them.

Example: 129th Infantry Regiment (United States)
https://en.wikipedia.org/wiki/129th_Infantry_Regiment_(United_States)

This article only lists books for further reading. Note there are no other sources included which would be useful to a researcher. The article is tagged at the top indicating citations and other information would be appreciated if a volunteer would like to add them to the article.

WorldCat
https://worldcat.org/

WorldCat is a searchable database of books, videos, films, and magazines for the entire world. Using WorldCat, often through your local library on-site or at home, you can search for books to request through inter-library loan if your library system does not have a specific book, magazine, or other media.

Search Tips: WorldCat is easy to use, and the entry form is straightforward. If a book does not appear, which you believe should appear, try adding or removing search terms until it is found.

Military Branch-Specific Resources

Air Force Historical Research Agency (AFHRA)
http://www.afhra.af.mil/

The Air Force Historical Research Agency maintains records for the Army Air Corps activities during the war. Records include Missing Air Crew Reports, Accident Reports, unit histories, and mission reports.

Naval History and Heritage Command
https://www.history.navy.mil/

The research center and website have been undergoing renovation, but are due to be open and full of information in 2015.

U.S. Army Center of Military History
http://history.army.mil/

The resources on this website are vast and include the Army Green Books, which detail the war and all the working parts; histories, research resources, and digitized materials.

United States Marine Corps History Division
https://www.usmcu.edu/historydivision

Visit their Frequently Requested page for information on records and resources. There is also information for units, digitized books, and research assistance.

Social Media Options

Facebook
http://facebook.com

Social media options abound for World War II research. Facebook hosts many groups and pages dedicated to units, divisions, reenactment groups, all service branches, and cemeteries. Use the search box to look for a specific group. For example, try 100th on its own. The search results appear for 100th Bomb Group, 100th Infantry Division, and so forth. Each page or group is named differently, so you need to explore to see if they are really what you seek.

Individuals on Facebook are also a great resource. As you interact with people in the World War II groups, friend some and share information. In Europe, the World War II historians, re-enactors, collectors, grave adopters, authors, and others have a great network. I have never encountered anyone who wasn't willing to offer some sort of help in the last several years when I have asked questions.

LinkedIn
http://linkedin.com

LinkedIn has professional groups for aspects of World War II and you can locate professionals doing specific research there. Just as on Facebook, try to connect with other professionals researching or writing about the unit you are interested in learning more about.

Twitter
http://twitter.com

Information and discussions are seen on Twitter through tweets. A tweet is a status update by a user in 140 characters or less. You can search two ways, using hashtags or a general user search. A hashtag is a keyword preceded by the # sign. For example, look for tweets about the 100th Bomb Group. First, try searching for #WWII. Many tweets will appear to choose from, so let's narrow it down and try #100 and see what appears. Does anything resembling #100thbg or #100thbombgroup or similar appear? If not, try a user search without the hashtag.

To search for a user, enter 100th in the search box. What appears? 100th Bomb Group @100thBG, 100th Bomb Group @100thBombGroup, etc. Explore those users to see who they are. You will find a couple are restaurants, and at least one is actually tweeting about the 100th Bomb Group during the war. You can converse with users on Twitter by replying to a tweet. Just remember to keep it 140 characters or less or break your question into multiple tweets.

Instagram
https://www.instagram.com/?hl=en

Instagram is an online forum where users post photographs or short videos of things in which they are interested. Many World War II researchers have joined Instagram in the last couple of years and post World War II-era photographs and current day photographs.

Blogs

A blog is a regularly updated website, typically one run by an individual or small group, that is written in an informal or conversational style. Blogs contain posts, or short articles, about a specific topic around which the website was created. Many family historians have blogs on which they document their family's history.

You may even find blogs contained within a unit website and the posts focus on aspects of that unit's history, military records, veteran stories, or reunion information.

Search for "WWII Blogs," "100th Bomb Group Blog," "World War II Blogs" or specific branches plus the word blog. Try a combination of things to locate what you are looking for. Search results will include any website or social media site with those keywords, not just blogs.

Pinterest
http://pinterest.com/

Finally, Pinterest is another popular social media site which functions as a bulletin board on which you can "pin" photos linked to articles or blog posts. The member boards are searchable, in the same ways you would search for blogs or specific units on Facebook. You can create your own boards, follow members who have boards in which you are interested, and pin things from other websites.

What's Next?

Have you gathered as much information as possible based on online sources? Have you exhausted the resources easily available to you offline? Would you like to learn how to obtain military records? Would you like to learn more about the OMPF, the IDPF, how to navigate company and unit level records, view record examples, and understand the information each record contains?

If you answered yes, then you are ready to learn the research strategy and move into ordering military records and finding the answers to your questions. To help you, the World War II Research and Writing Center is rolling out branch specific courses and books on our education website WWII Education (http://wwiieducation.com).

You can find everything currently available, listed in the Store. Please sign up for our free newsletter to stay informed of new class offerings, webinars, live events, and new books.

Researching your service member's history can be complex. The World War II Research and Writing Center provides expert research experts to tackle your most challenging research questions. Contact us at info@wwiirwc.com for project details and costs. We have researchers on-site at the National Archives facilities to obtain materials within a few weeks, and a network of researchers and tour guides around the globe.

Appendix

Military Records Online

The purpose of this list is to provide ideas on where to find information available in digital format. This is not an exhaustive list of online resources. Websites change daily and it is impossible to track every change to records online.

____ American Battle Monuments Commission (ABMC)
____ Ancestry.com
____ Army Enlistment Database
____ Army Registers, 1798-1969
____ BillionGraves entries
____ FindAGrave entries
____ Fold3.com
____ Foreign Burial of American War Dead
____ Headstone Applications, 1925-1963
____ Honolulu, Hawaii, National Memorial Cemetery of the Pacific (Punchbowl), 1941-2011
____ Individual State Casualty Lists
____ Marine Corps Muster Rolls
____ Medal of Honor Recipients, 1863-2013
____ Missing Air Crew Reports
____ Newspapers: Memorial Notices and Obituaries, Lists of POW, MIA, KIA
____ New York National Guard Records
____ Pearl Harbor Muster Rolls
____ Stars and Stripes Newspapers
____ State Department Records - France
____ State Department Records - Russia
____ U.S. American Red Cross Nurse Files
____ U.S., Headstone and Interment Records for U.S. Military Cemeteries on Foreign Soil, 1942-1949
____ U.S. Merchant Marine Applications for License of Officers, 1914-1949
____ U.S. Navy Cruise Books
____ U.S. Rosters of WWII Dead
____ U.S., War Department, Press Releases and Related Records, 1942-1945
____ U.S. WWII Jewish Servicemen Cards
____ Utah, Death & Military Death Certificates, 1904-1961

_____ Veterans Affairs BIRLS Death File
_____ WWII Bonus Case Files (Veteran Compensation Applications)
_____ WWII Cadet Nursing Corps Card Files
_____ WWII European Theater Army Records
_____ WWII Foreign Military Studies
_____ WWII JAG Case Files
_____ WWII Missing In Action or Lost at Sea
_____ WWII Naval Press Clippings
_____ WWII Navy, Marine Corps and Coast Guard Casualties, 1941-1945
_____ WWII Prisoners of the Japanese, 1941-1945
_____ World War II Prisoners of War Data File, 12/7/1941 - 11/19/1946
_____ WWII Submarine Patrol Reports
_____ WWII U.S. Navy Muster Rolls
_____ Young American Patriots Military Yearbooks

About the Author

About The Author

Global Coordinator of the World War II Research and Writing Center

Jennifer Holik is an acclaimed author, researcher, educator, empath, medium, and healer dedicated to uncovering WWII history by piecing together biographical stories of soldiers' lives that have never been told before. She has a rare talent for telling WWII stories in an emotive way: lending a 'new voice' to this period of history by shedding light on a soldier's relationships with their loved ones, family and friends. In portraying the human side of warfare she reveals in a poignant, heartfelt, original way what it was really like to have a son, brother, best friend or spouse go off to fight this incredulous war and risk making the ultimate sacrifice for liberty and freedom. Through the research and stories, Jennifer provides healing of the past for clients and those who have already left us.

Based in Chicago, Illinois and Amstelveen, Netherlands, her unique talent and capacities bringing to life a soldier's story by new research techniques provides a rare glimpse into a soldier's personal life. The facts she unravels about his web of relationships provide family members a chance to revisit their soldier's never-told-before story and 'personal war journey' in a new way. In doing so, it allows family members to live their soldiers' pain and glory – memorialize their stories through writing – ultimately, serving as a tribute to their sacrifice; and a testament to our great country.

Hear her story in a brief film about her work. You can view it on her website at http://wwiirwc.com.

www.ingramcontent.com/pod-product-compliance
Lightning Source LLC
Chambersburg PA
CBHW060700280326
41933CB00012B/2258